ABCs
of Over-The-Counter Medications
Chemistry 101

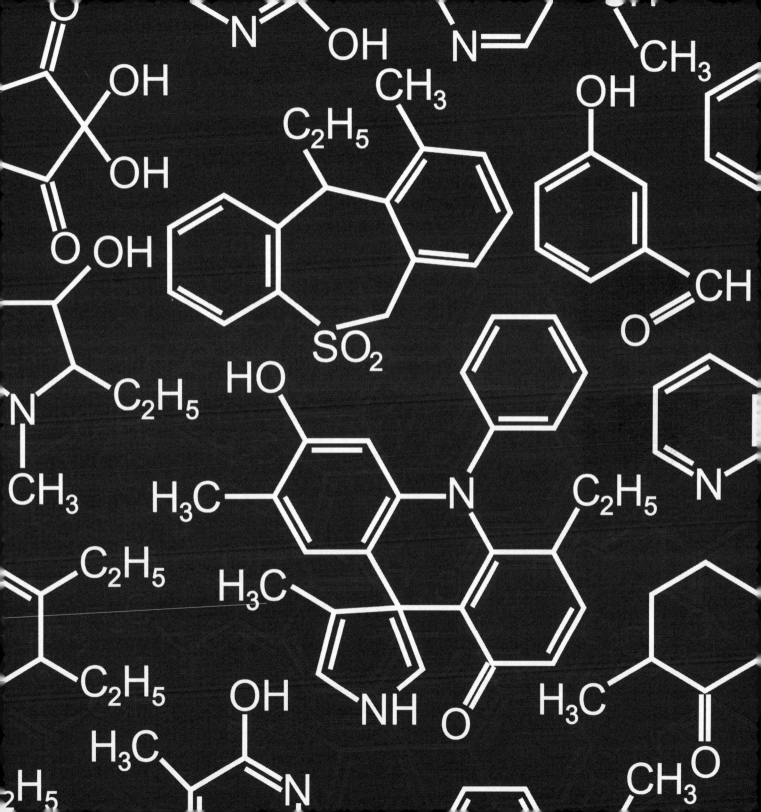

ABCs

of Over-The-Counter Medications
Chemistry 101

Have you ever wondered what over-the-counter (OTC) drugs look like in their chemical structure forms? This book takes us from A to Z to learn the commonly seen OTC products with a flavor of chemistry for all ages. We are accommodating our youngest audience with a black-and-white color scheme.

HELLO
my name is

FUTURE PHARMD

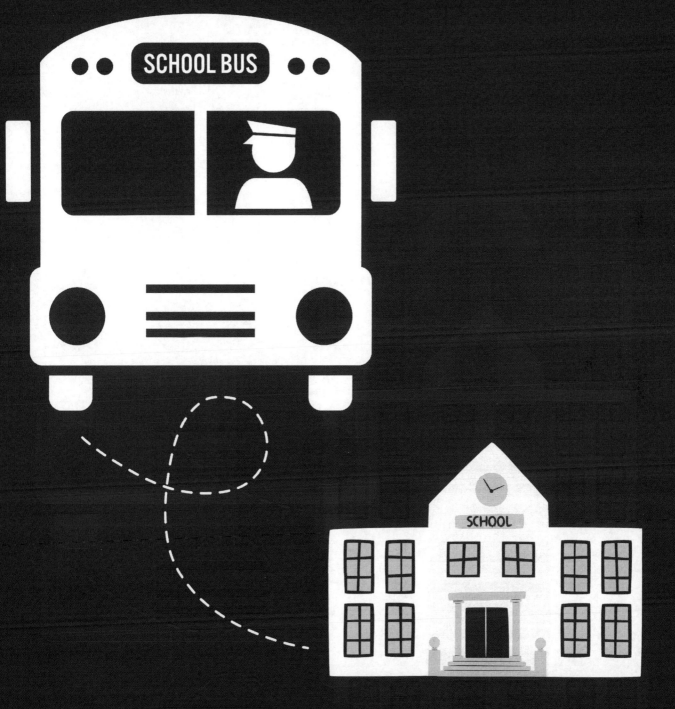

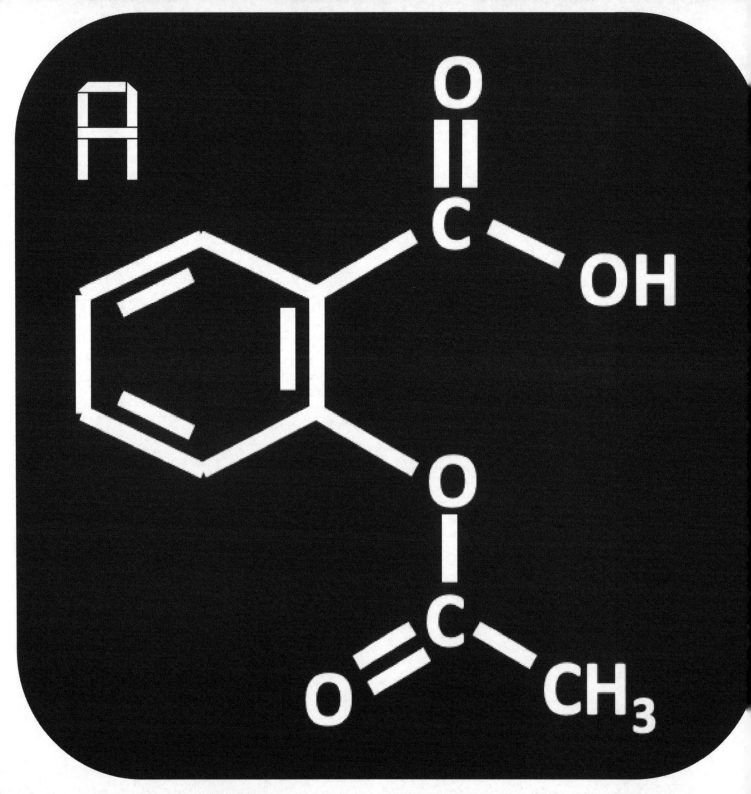

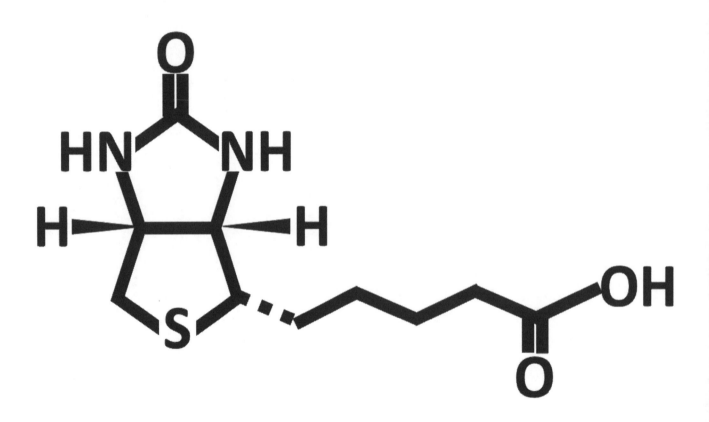

B is for
Biotin

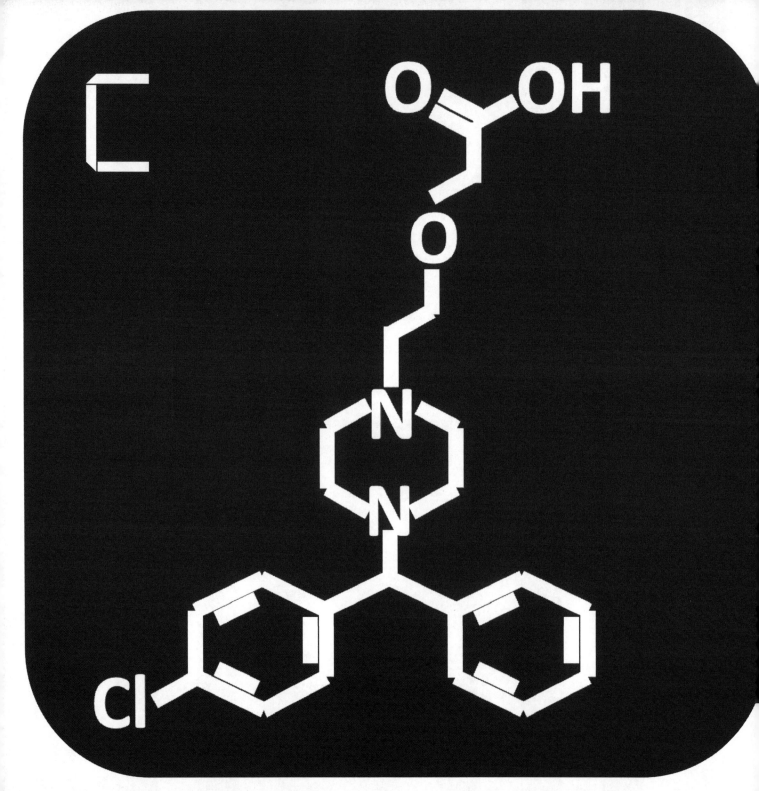

C is for

Cetirizine

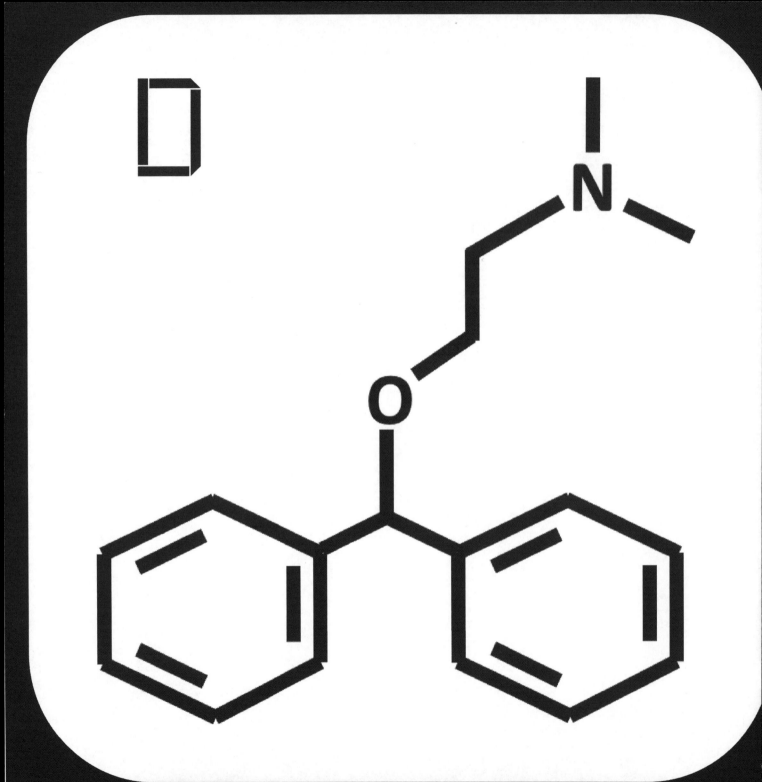

D is for

Diphenhydramine

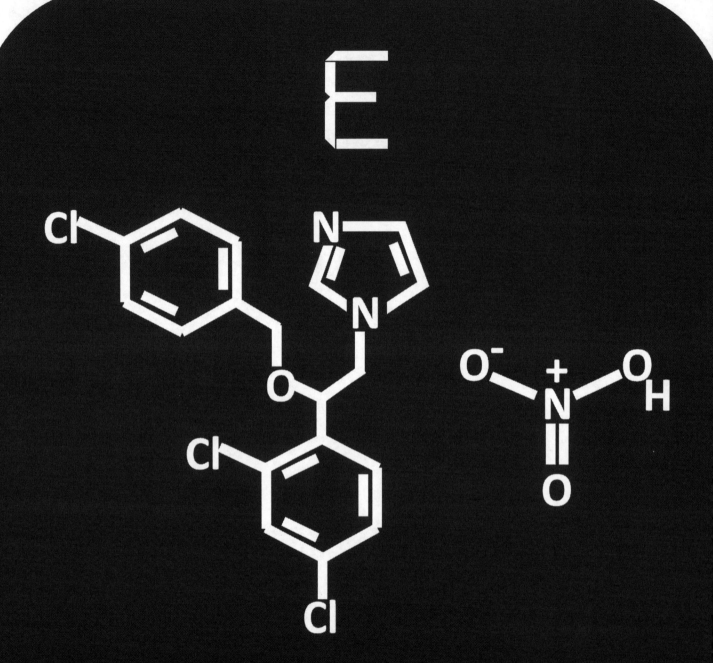

E is for
Econazole
Nitrate Cream

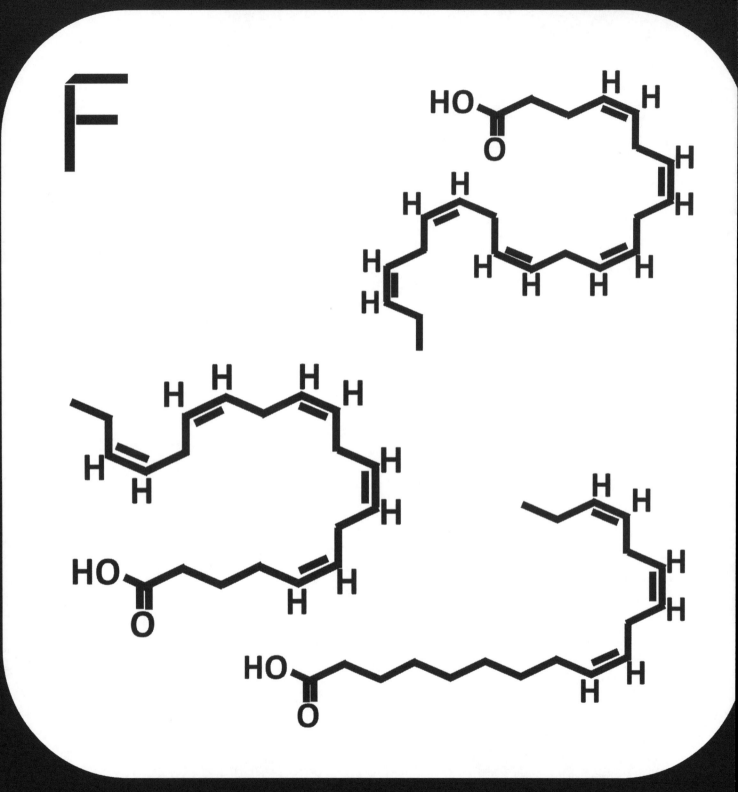

F is for
Fish Oil

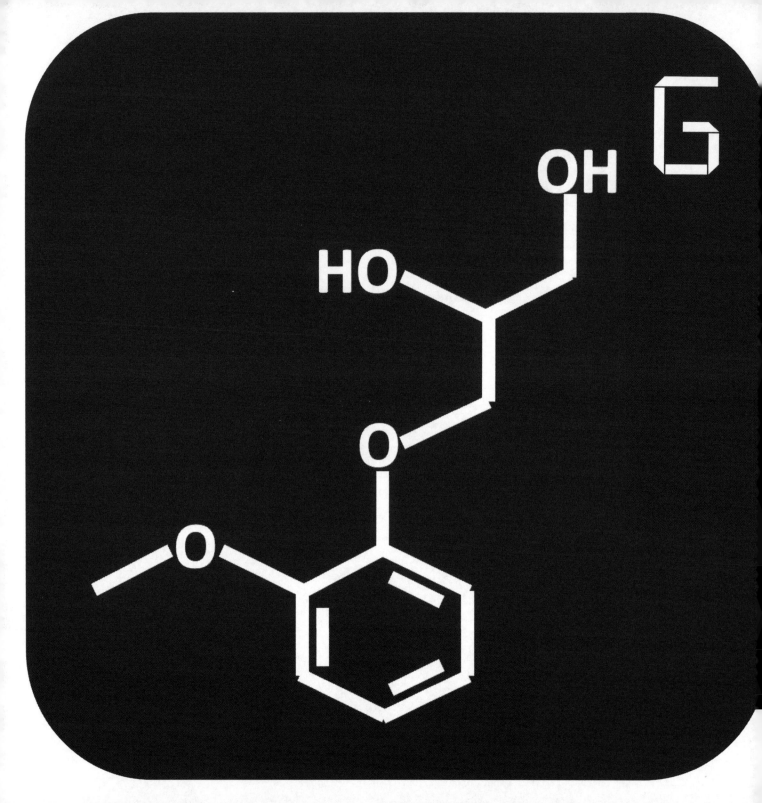

G is for Guaifenesin

H

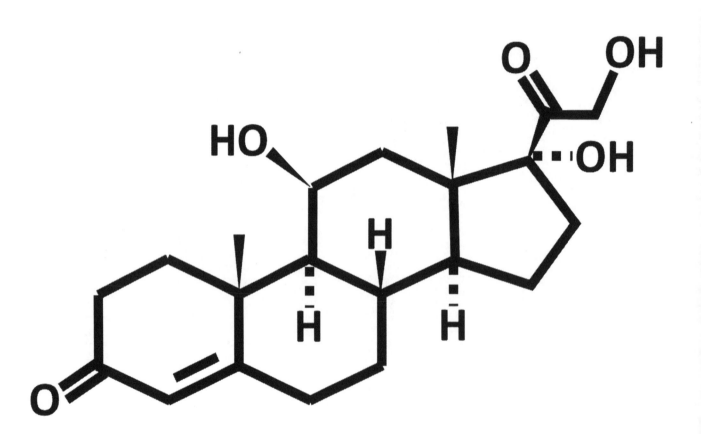

H is for
Hydrocortisone
Cream

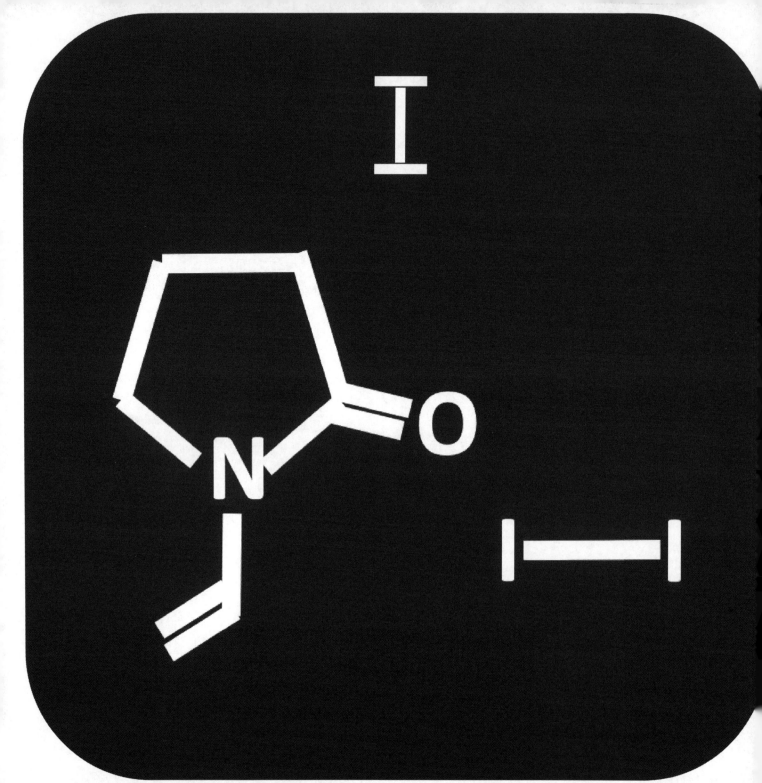

I is for

Iodine Povidone

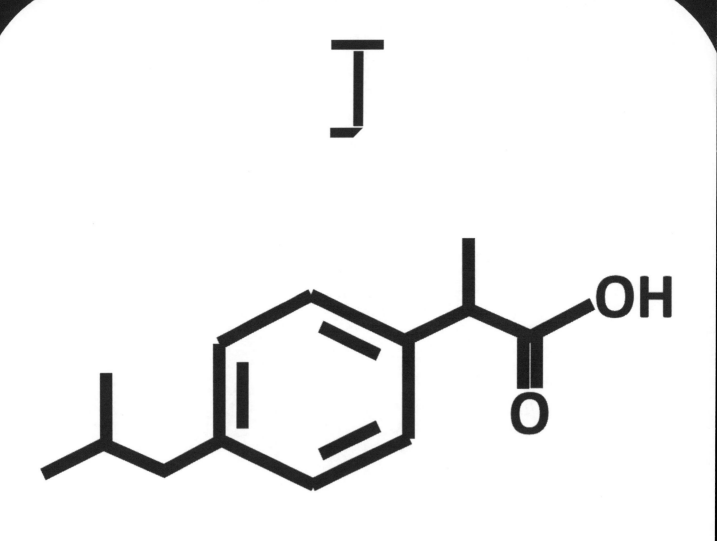

J is for

Junior Strength

Ibuprofen

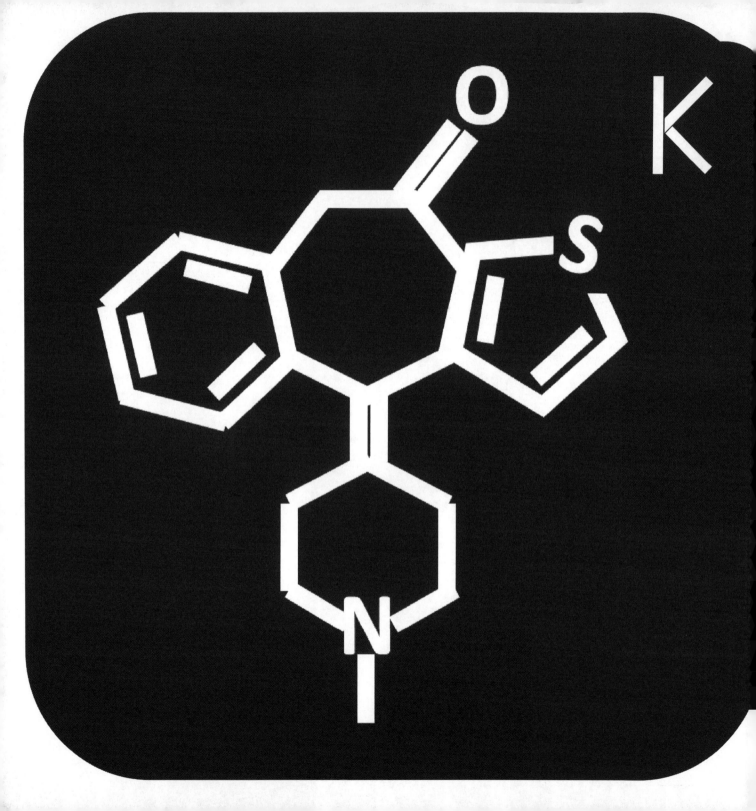

K is for

Ketotifen Eye Drops

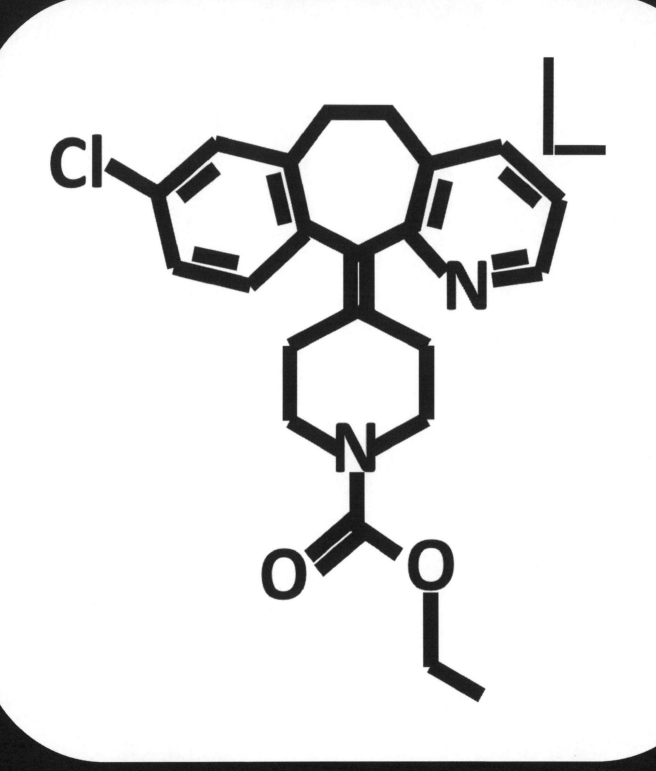

L is for
Loratadine

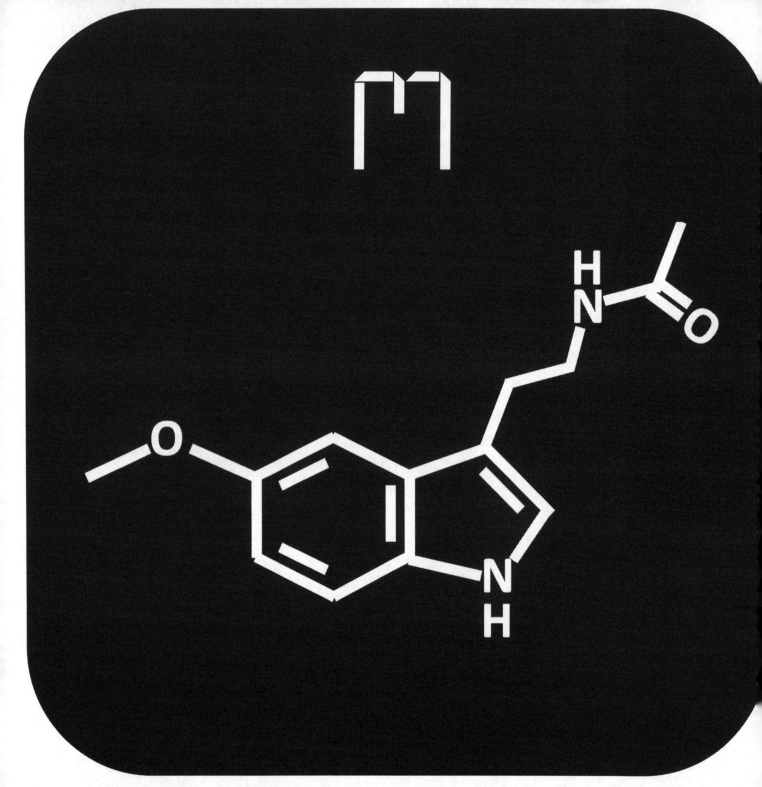

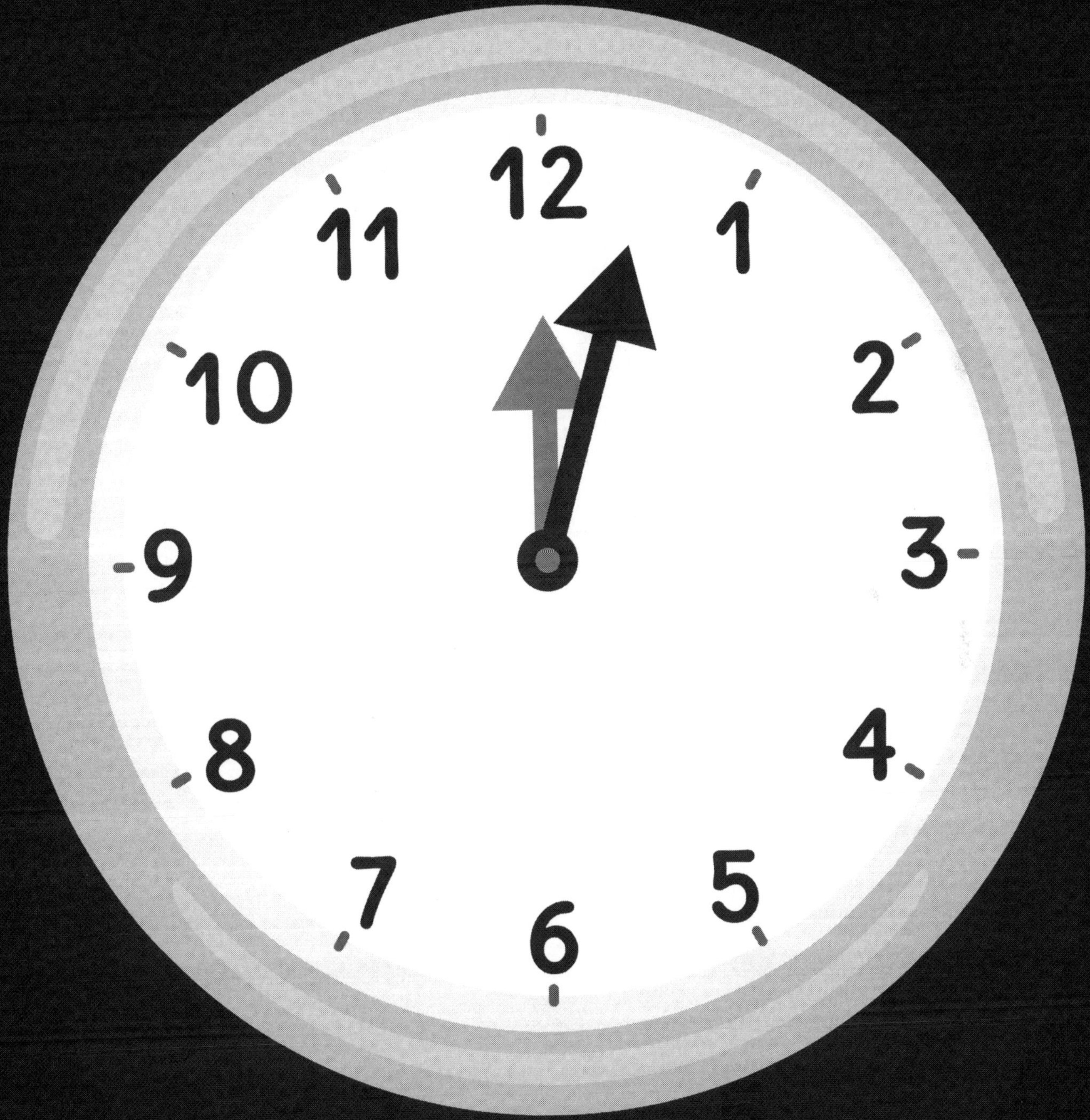
12
1
2
3
4
5
6
7
8
9
10
11

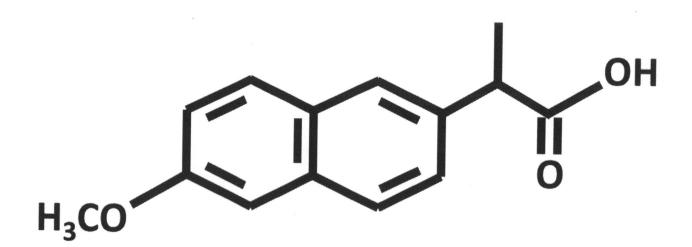

N is for

Naproxen

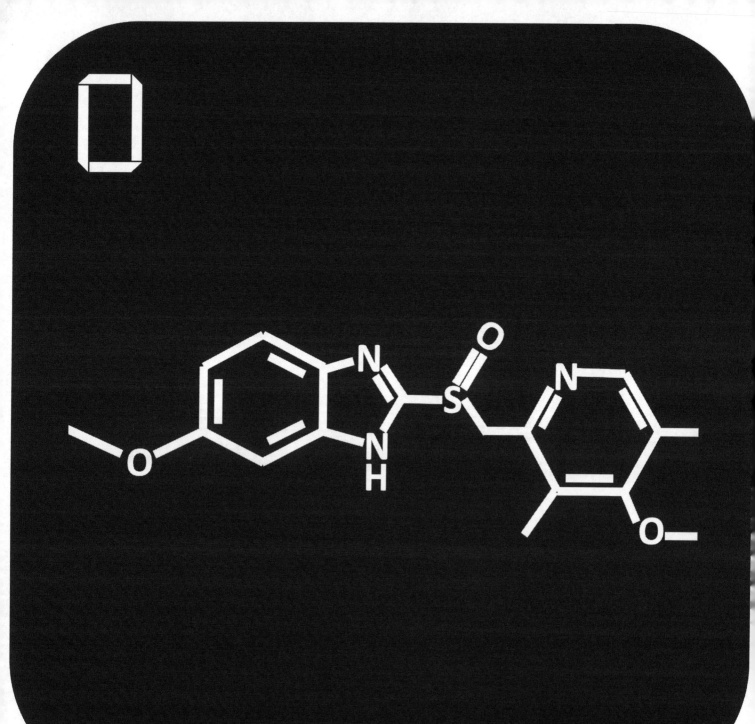

O is for

Omeprazole

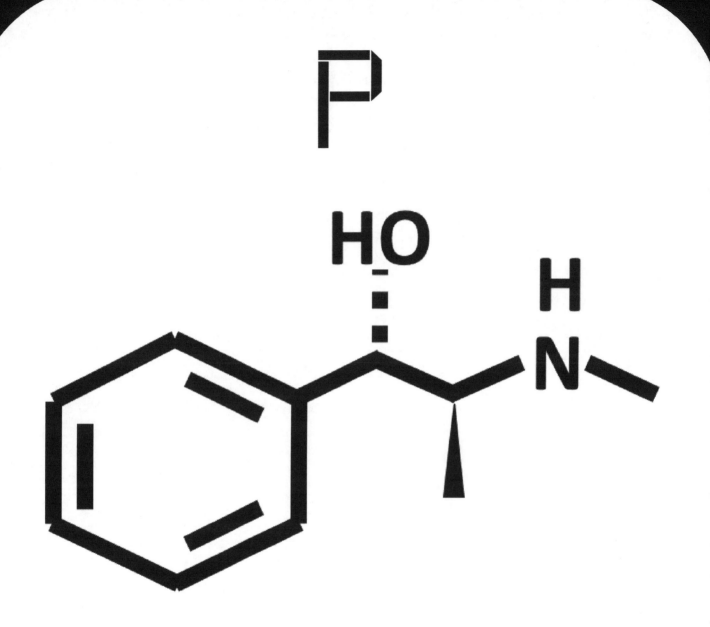

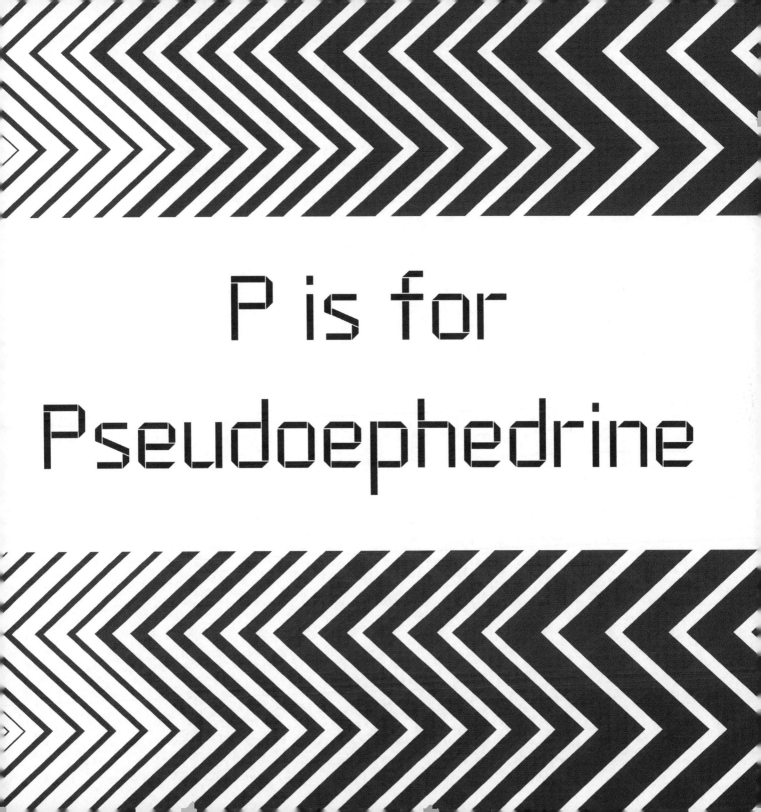

P is for

Pseudoephedrine

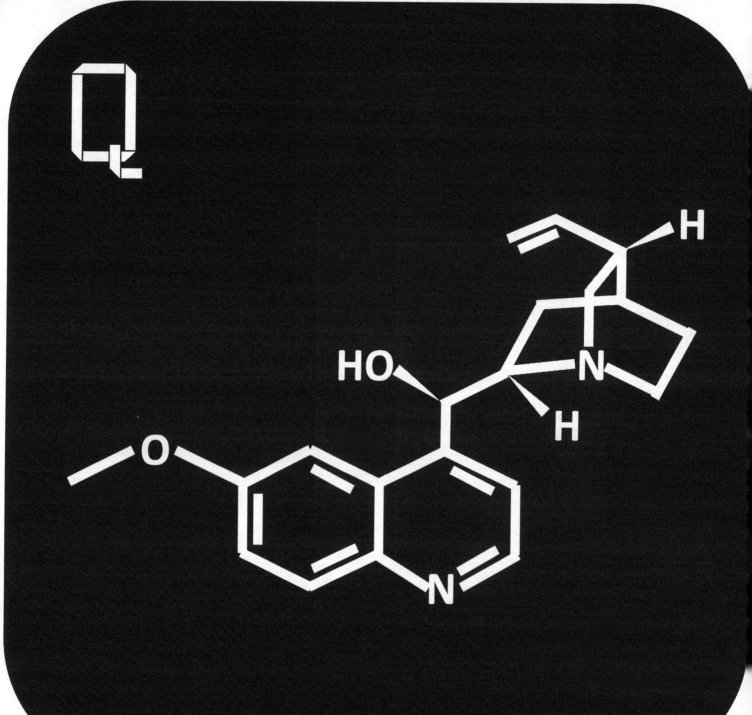

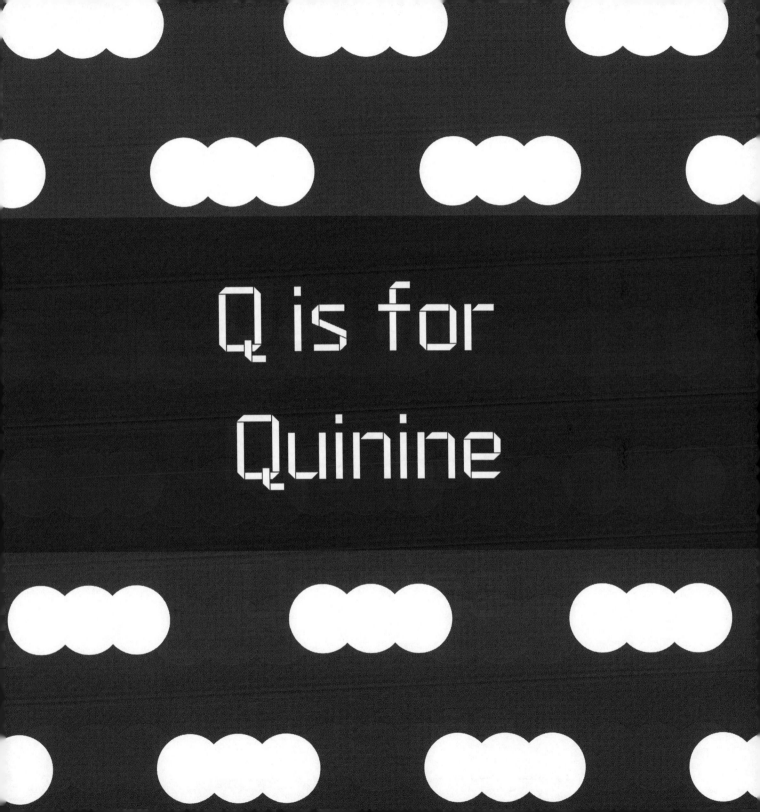

Q is for

Quinine

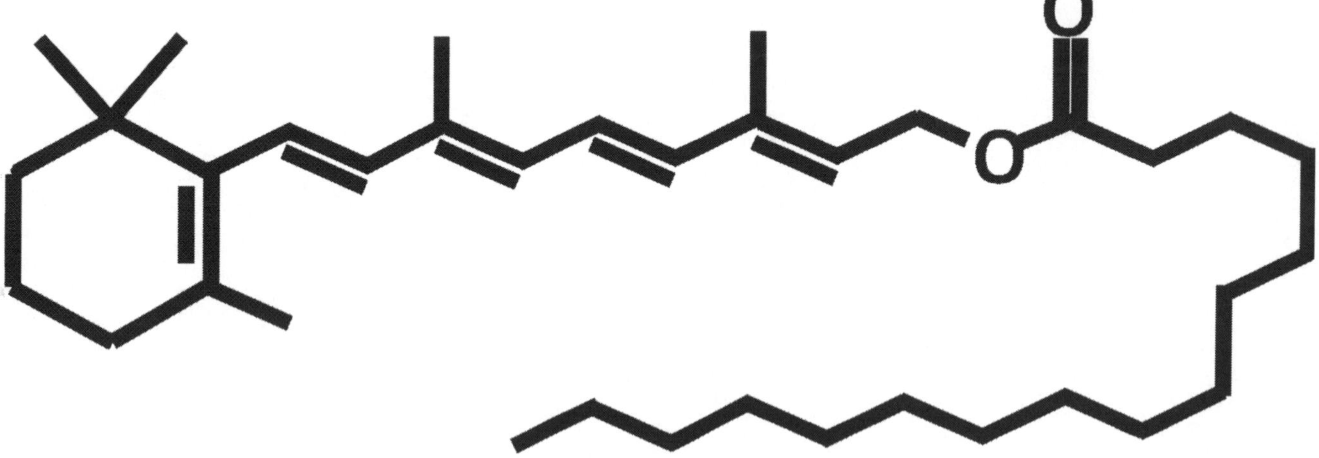
O
O

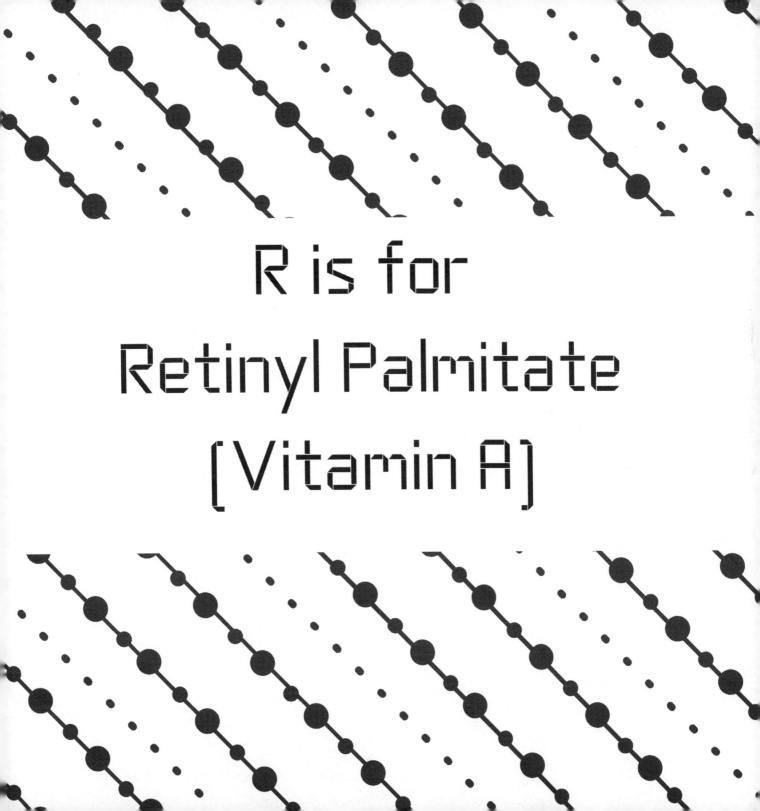

R is for
Retinyl Palmitate
(Vitamin A)

S

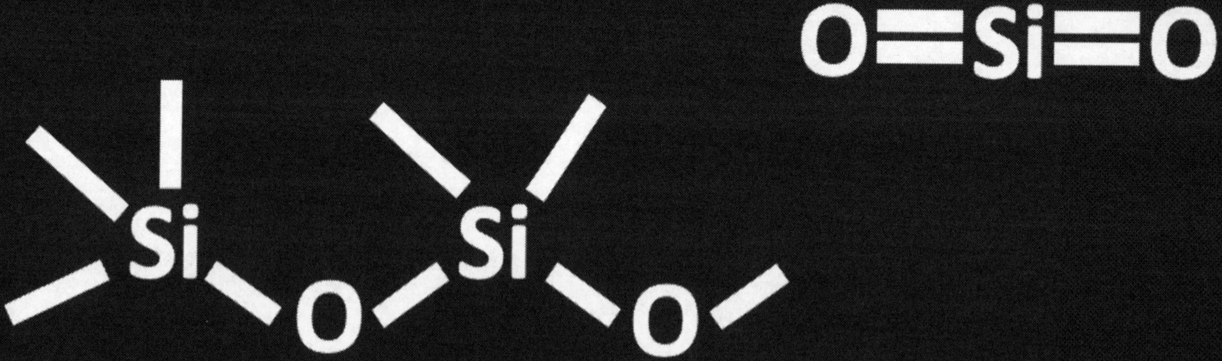

S is for

Simethicone

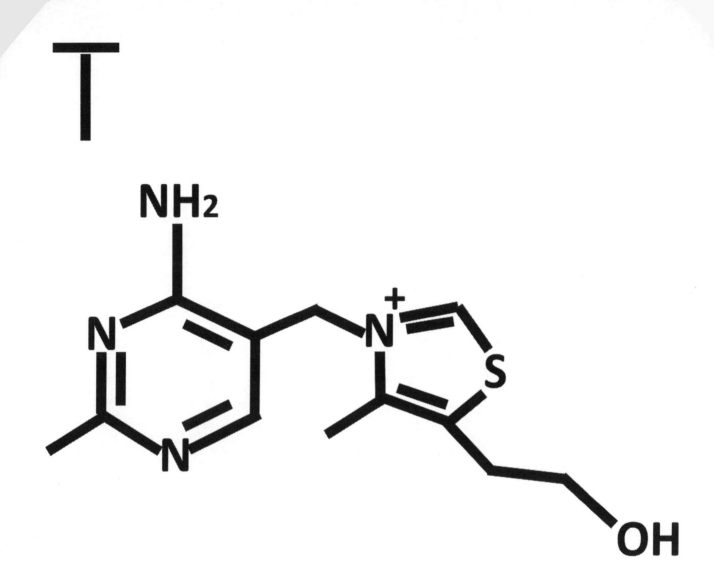

T is for

Thiamine

[Vitamin B1]

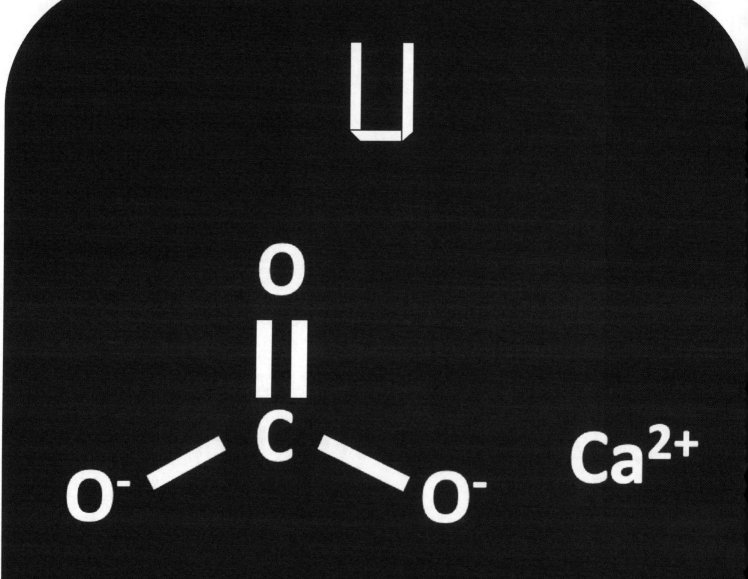

U is for

Ultra strength

antacid [calcium

carbonate]

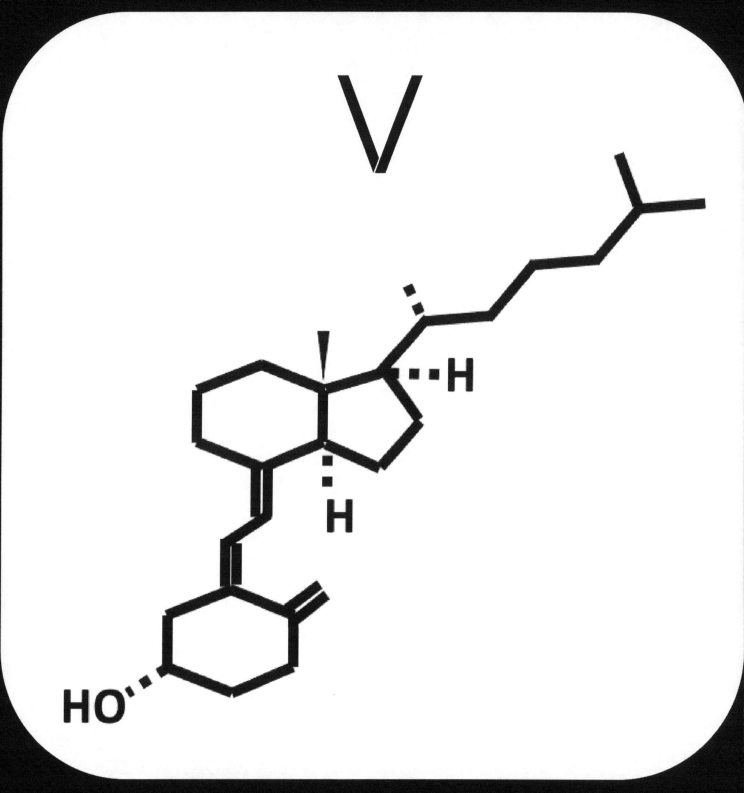

W

$$Na^+ \cdots \cdots \cdots Cl^-$$

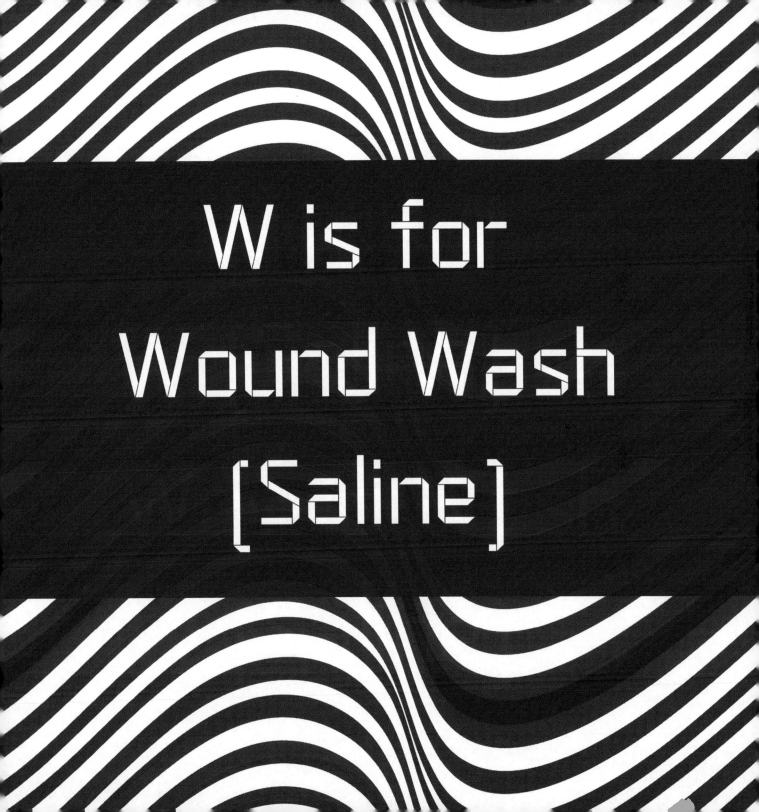

W is for Wound Wash [Saline]

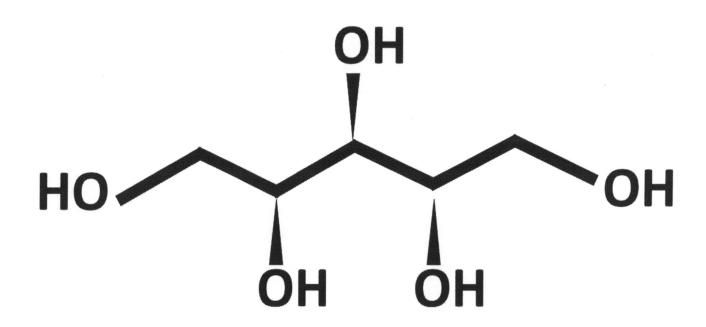

X is for

Xylitol

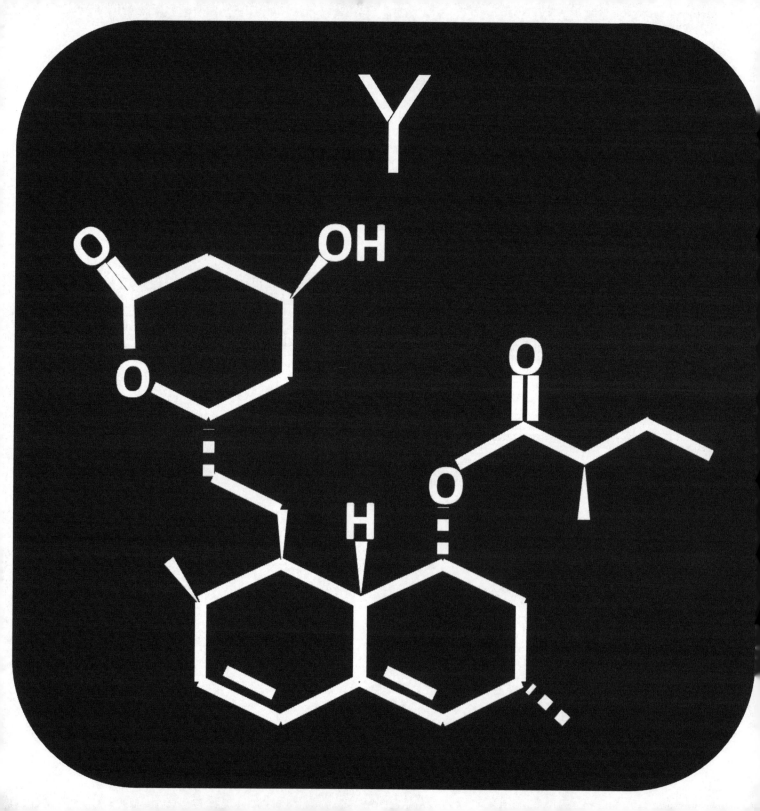

Y is for Yeast Rice [Red]

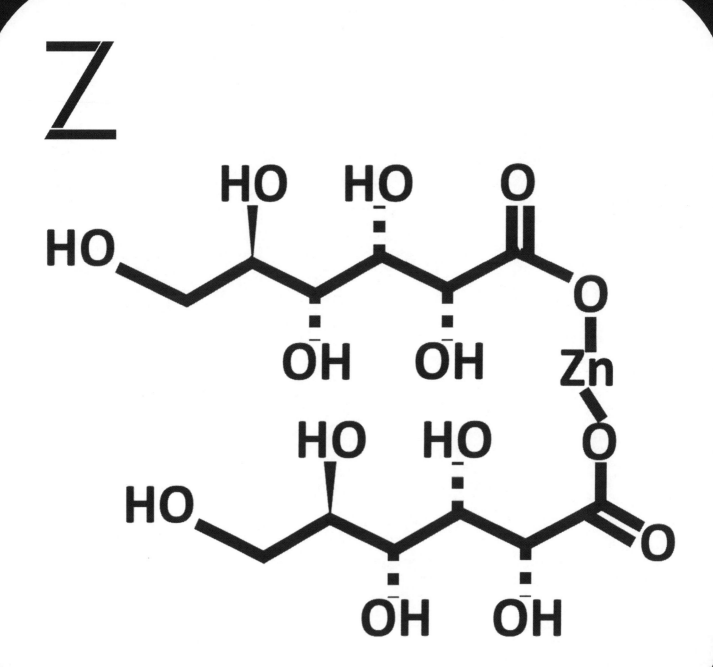

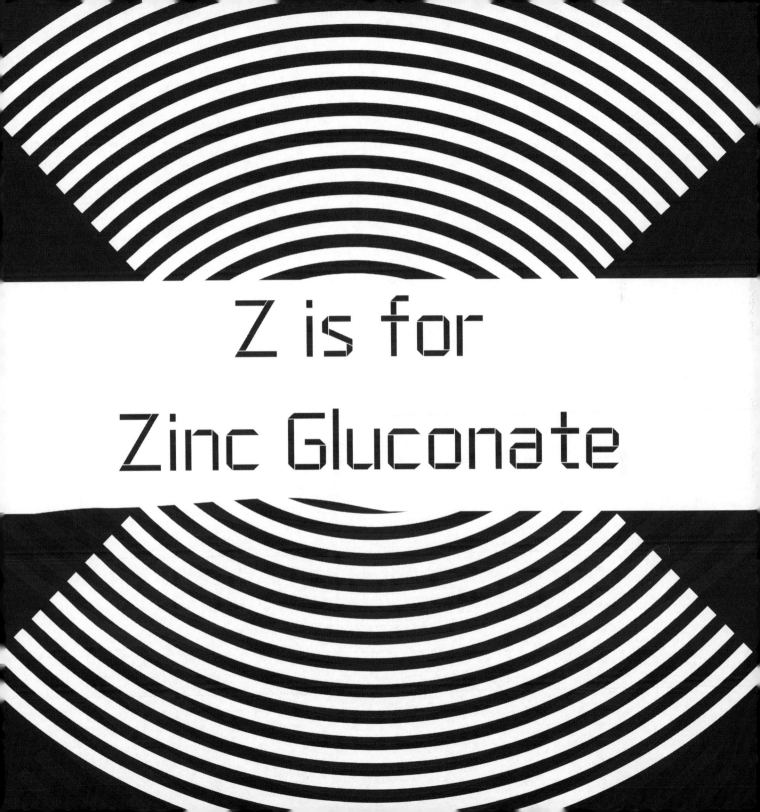

Z is for

Zinc Gluconate

Author Team

Dr. Amatullah is a pharmacist, informaticist, and artist. She knows the ins and outs of specialty pharmacy during the day and shifts to taking care of her toddler son and newborn daughter. Her life does not have a dull moment, and now she is even more complete being an author.

Dr. Li is a practicing clinician specializing in chronic disease management with administrative responsibilities and conducting clinical research. She is inspired by her toddler daughter every day and cannot wait to have an influence on the next generation of young scientists.

Dr. Tsapepas is a pharmacist with years of clinical experience who now focuses on quality, education and healthcare research. She is the best Thea Deda to her nephews and nieces who keep her entertained. She would like to extend learning excitement to all.

The three close friends have partnered on multiple projects over the years. Their latest adventure is to create fun educational materials and foster critical thinking in children.

THE

END!

Children's Books

Daskalos Lab

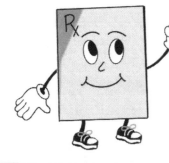

Scan the QR code to visit our social media page and order your book(s)!

Printed by Amazon Italia Logistica S.r.l.
Torrazza Piemonte (TO), Italy

54715854R00042